# LETTERS AND NUMBERS

SECOND EDITION

JOHN T. TORTORA

authorHOUSE®

*AuthorHouse™*
*1663 Liberty Drive*
*Bloomington, IN 47403*
*www.authorhouse.com*
*Phone: 833-262-8899*



*Published by AuthorHouse 10/11/2022*

*ISBN: 978-1-6655-7313-9 (sc)*
*ISBN: 978-1-6655-7323-8 (e)*

*Print information available on the last page.*

# DEDICATION

This book is written and dedicated in love and memory of my Mom Judy Ann Savage and Nana Ann Marchitelli.

John

I have thoroughly enjoyed assisting John in editing this book. I am proud of you, John, and I know you Mom and Nana would be very proud of you too. God has created you uniquely, not like anyone else, but exactly like you. Live for Jesus and you will be blessed. Whenever you get down, always remember "I will praise thee; for I am fearfully and wonderfully made: marvellous are thy works; and that my soul knoweth right well." Psalm 139:14

Krys

# PREFACE

The concept of this book is to convert numbers to letters and letters to numbers. A=1, B=2, C=3 through Z=26.

By converting the numbers and letters the reader will make words and solve mathematical problems.

For example:

- Make a word: 3, 1, 2 = CAB
- Addition: A+B=3

The mathematical problems become more complex as the units progress to include multiplying letters then dividing a number with the remainder number to equal a letter.

There are so many different uses and applications for this book. Here are just a few:

- Children
- Home school
- Tutoring
- Schools
- Behavioral Health tool
- Adults
- Cognitive Stimulation Therapy
- Fun puzzle problems similar to logic, crossword or sudoku

# CONTENTS

# KEY

Instructions: Use this key to solve the problems throughout the entire book.

A = 1
B = 2
C = 3
D = 4
E = 5
F = 6
G = 7
H = 8
i = 9
J = 10
K = 11
L = 12
M = 13

N = 14
O = 15
P = 16
Q = 17
R = 18
S = 19
T = 20
U = 21
V = 22
W= 23
X = 24
Y = 25
Z = 26

UNIT 1

# NUMBERS TO WORDS

# NUMBERS TO WORDS - 1

1. 7 9 18 12 19
2. 11 9 14 4
3. 2 15 25 19
4. 17 21 5 5 14 19
5. 13 1 14 1 7 5
6. 16 1 18 5 14 20 19
7. 6 18 9 4 7 5
8. 22 5 8 9 3 12 5

# NUMBERS TO WORDS - 2

1. 6 9 7 8 20 5 18
2. 3 1 18 19
3. 16 12 1 14 5 19
4. 1 4 15 12 5 19 3 5 14 20
5. 19 20 21 4 5 14 20
6. 7 18 1 14 4 11 9 4
7. 11 9 14 7
8. 17 21 5 5 14

# NUMBERS TO WORDS - 3

1. 4 5 1 20 8
2. 8 5 1 20
3. 12 5 6 20
4. 13 15 14 5 25
5. 16 15 15 18
6. 12 15 1 14
7. 6 1 13 9 12 25

# NUMBERS TO WORDS - 4

ALL END WITH AN "S"

1. 2 1 20 19
2. 11 9 4 19
3. 12 5 7 19
4. 16 1 18 5 14 20 19
5. 19 20 9 3 11 19
6. 18 5 16 20 9 12 5 19
7. 19 9 2 12 9 14 7 19

# NUMBERS TO WORDS - 5

ALL END WITH "ED"

1. 13 1 18 18 9 5 4
2. 16 12 1 14 20 5 4
3. 7 18 1 14 20 5 4
4. 3 15 22 5 18 5 4
5. 23 15 18 18 9 5 4
6. 20 18 9 5 4
7. 15 16 5 14 5 4

# NUMBERS TO WORDS - 6

## STATES

1. 3 15 12 15 18 1 4 15
2. 3 15 14 14 5 3 20 9 3 21 20
3. 23 25 15 13 9 14 7
4. 13 9 3 8 9 7 1 14
5. 7 5 15 18 7 9 1
6. 1 12 1 2 1 13 1
7. 14 5 2 18 1 19 11 1

# NUMBERS TO WORDS - 7

BOYS NAMES

1. 2 18 9 1 14
2. 2 18 1 4
3. 6 18 5 4
4. 10 15 8 14
5. 19 20 5 22 5
6. 4 1 22 9 4
7. 13 9 11 5
8. 3 8 18 9 19

# NUMBERS TO WORDS - 8

GIRLS NAMES

1. 12 9 19 1
2. 10 5 14
3. 16 1 13
4. 4 5 2 2 9 5
5. 2 18 9 4 7 5 20
6. 19 20 5 16 8 1 14 9 5
7. 10 5 19 19 9 3 1

UNIT 2

# ADDITION

# ADDITION - 1

## ADDING NUMBERS TO EQUAL A LETTER
## NO LETTER WILL BE THE SAME

1. 9+10
2. 6+15
3. 1+13
4. 22+2
5. 4+8
6. 11+5
7. 3+14

# ADDITION - 2

ADDING NUMBERS TO EQUAL A LETTER
THEN COMBINE THE LETTERS TO
MAKE A 7 LETTER WORD

1. 15+4
2. 2+1
3. 4+4
4. 11+4
5. 9+6
6. 6+6
7. 11+8

## ADDITION - 3

ADDING NUMBERS TO EQUAL A LETTER
THEN COMBINE THE LETTERS
TO MAKE AN 8 LETTER WORD

1. 10+10
2. 3+2
3. 0+1
4. 2+1
5. 4+4
6. 4+1
7. 10+8
8. 10+9

# ADDITION - 4

ADDING NUMBERS TO MAKE A WORD

1. 3+10 2+13 0+1 12+2 6+3 13+1 2+5
2. 4+2 1+4 11+2 0+1 10+2 1+4 16+3
3. 16+2 3+2 17+2 18+2 3+2 2+2
4. 16+3 15+5 17+4 0+4 1+4 12+2 18+2
5. 9+2 8+1 17+3 10+10 3+2 11+3
6. 11+9 6+3 4+3 4+1 16+2
7. 14+3 18+3 0+1 10+2 6+3 18+2 12+13

# ADDITION - 5

## ADDING LETTERS TO EQUAL A NUMBER

1. B + M + H + i
2. Q + T + S + K
3. A + M + D + G
4. S + U + M + O
5. X + W + Y + Z
6. Y + R + B + G
7. V + Q + i + W
8. Z + U + T + P
9. U + H + M + O
10. V + i + J + M

# ADDITION – 6

## ADDING LETTERS IN A WORD TO EQUAL A NUMBER

1. KITTEN =
2. TEACHER =
3. PARENT =
4. HOUSE =
5. COUNTRY =
6. STATE =
7. COUNTY =
8. ANIMAL =

UNIT 3

# SUBTRACTION

# SUBTRACTION - 1

SUBTRACTING NUMBERS TO GET A LETTER
NO LETTER WILL BE THE SAME

1. 36-28
2. 44-20
3. 54-34
4. 55-43
5. 60-58
6. 53-28
7. 61-60

# SUBTRACTION - 2

## SUBTRACTING NUMBERS TO MAKE A WORD

1. 5-2 4-3 21-7 6-2 24-12 15-10
2. 12-8 18-3 20-8 24-12 25-6
3. 17-5 11-2 23-4 22-2
4. 26-7 21-1 12-7 21-5 22-3
5. 12-8 11-6 6-5 16-10
6. 15-3 15-6 17-10 10-2 22-2 22-3
7. 15-4 14-5 7-3 24-5

# SUBTRACTION - 3

SUBRACTING LETTERS FROM NUMBERS
THE ANSWER IS A LETTER

1. 32 - J =

2. 29 - Y =

3. 27 - F =

4. 33 - V =

5. 30 - T =

6. 37 - S =

7. 31 - X =

8. 34 - S =

9. 35 - Q =

10. 40 - U =

UNIT 4

# ADDITION WITH SUBTRACTION

# ADDITION WITH SUBTRACTION - 1

## ADDING SINGLE DIGIT NUMBERS THEN SUBTRACTING SINGLE DIGIT NUMBERS TO EQUAL A LETTER

1.
```
   3
 + 5
-----
=
 - 4
-----
```

2.
```
   5
 + 8
-----
=
 - 3
-----
```

3.
```
   7
 + 9
-----
=
 - 6
-----
```

4.
```
   9
 + 9
-----
=
 - 7
-----
```

5.
```
   8
 + 6
-----
=
 - 2
-----
```

6.
```
   7
 + 7
-----
=
 - 5
-----
```

7.
```
   4
 + 3
-----
=
 - 1
-----
```

8.
```
   6
 + 6
-----
=
 - 8
-----
```

# ADDITION WITH SUBTRACTION - 2

ADDING NUMBERS THEN
SUBTRACTING NUMBERS TO EQUAL A LETTER

1.
$$\begin{array}{r} 10 \\ +\ 4 \\ \hline = \quad \\ -\ 3 \\ \hline \end{array}$$

2.
$$\begin{array}{r} 13 \\ +\ 9 \\ \hline = \quad \\ -\ 5 \\ \hline \end{array}$$

3.
$$\begin{array}{r} 18 \\ +\ 7 \\ \hline = \quad \\ -\ 5 \\ \hline \end{array}$$

4.
$$\begin{array}{r} 24 \\ +\ 9 \\ \hline = \quad \\ -\ 7 \\ \hline \end{array}$$

5.
$$\begin{array}{r} 21 \\ +\ 7 \\ \hline = \quad \\ -\ 8 \\ \hline \end{array}$$

6.
$$\begin{array}{r} 16 \\ +\ 11 \\ \hline = \quad \\ -\ 4 \\ \hline \end{array}$$

7.
$$\begin{array}{r} 15 \\ +\ 14 \\ \hline = \quad \\ -\ 13 \\ \hline \end{array}$$

# ADDITION WITH SUBTRACTION - 3

ADDING NUMBERS THEN SUBTRACTING NUMBERS
THE ANSWER IS A LETTER

1.
```
   31
 + 29
 ----
 =
 - 37
 ----
```

2.
```
   34
 + 26
 ----
 =
 - 38
 ----
```

3.
```
   38
 + 29
 ----
 =
 - 47
 ----
```

4.
```
   23
  +21
 ----
 =
 - 35
 ----
```

5.
```
   33
 + 31
 ----
 =
 - 51
 ----
```

6.
```
   14
 + 12
 ----
 =
 - 18
 ----
```

7.
```
   47
 + 42
 ----
 =
 - 82
 ----
```

8.
```
   49
 + 32
 ----
 =
 - 77
 ----
```

# ADDITION WITH SUBTRACTION - 4

ADDING NUMBERS THEN SUBTRACTING NUMBERS
THE ANSWER IS A LETTER

1.
 36
+ 33
=
- 56

2.
 47
+ 23
=
- 49

3.
 39
+ 17
=
- 44

4.
 58
+ 28
=
- 67

5.
 64
+ 43
=
- 97

6.
 59
+ 31
=
- 79

7.
 64
+ 22
=
- 84

8.
 55
+ 21
=
- 67

# ADDITION WITH SUBTRACTION - 5

ADDING NUMBERS THEN SUBRACTING LETTERS

1.
```
  25
+ 22
----
=
 - P
----
```

2.
```
  28
+ 24
----
=
 - H
----
```

3.
```
  29
+ 21
----
=
 - K
----
```

4.
```
  32
+ 30
----
=
 - R
----
```

5.
```
  37
+ 26
----
=
 - V
----
```

6.
```
  39
+ 35
----
=
 - Y
----
```

7.
```
  33
+ 28
----
=
 - B
----
```

8.
```
  40
+ 34
----
=
 - G
----
```

# ADDITION WITH SUBTRACTION - 6

## ADDING LETTERS THEN SUBTRACTING NUMBERS

1.
```
   W
 + G
 ___
 =
 - 7
 ___
```

4.
```
   S
 + Q
 ___
 =
 - 5
 ___
```

7.
```
   S
 + F
 ___
 =
 - 4
 ___
```

2.
```
   N
 + H
 ___
 =
 - 10
 ___
```

5.
```
   X
 + K
 ___
 =
 - 12
 ___
```

8.
```
   V
 + M
 ___
 =
 - 9
 ___
```

3.
```
   P
 + O
 ___
 =
 - 8
 ___
```

6.
```
   Z
 + i
 ___
 =
 - 3
 ___
```

# ADDITION WITH SUBTRACTION – 7

## ADDING NUMBERS AND LETTERS THEN SUBTRACTING WITH LETTERS

1.

28
\+ Z
=
\- E

2.

31
\+ R
=
\- G

3.

33
\+ Y
=
\- D

4.

37
\+ S
=
\- K

5.

39
\+ M
=
\- V

6.

41
\+ i
=
\- Y

7.

44
\+ Q
=
\- T

8.

42
\+ N
=
\- W

# ADDITION WITH SUBTRACTION - 8

ADDING LETTERS
THEN SUBTRACTING LETTERS
TO EQUAL A NUMBER

1. T + M - K =
2. Q + L - N =
3. Z + X - U =
4. J + I - D =
5. V + W - H =
6. P + O - B =
7. R + G - C =
8. U + J - E =

# ADDITION WITH SUBTRACTION - 9

ADDING THE LETTERS IN A WORD
THEN SUBTRACTING A NUMBER
TO EQUAL A LETTER

1. HIGH – 12 =
2. CALM – 11 =
3. KIND – 19 =
4. LOVE – 32 =
5. JOBS – 39 =
6. BRAVE – 44 =
7. SWEET – 63 =
8. SCHOOL – 61 =

UNIT 5

# MULTIPLICATION

# MULTIPLYING – 1

## NUMBERS MULTIPLYING LETTERS
## NUMBERS 20 AND UNDER

1. $\begin{array}{r} 19 \\ \times\ Y \\ \hline \end{array}$

2. $\begin{array}{r} 17 \\ \times\ F \\ \hline \end{array}$

3. $\begin{array}{r} 16 \\ \times\ Z \\ \hline \end{array}$

4. $\begin{array}{r} 20 \\ \times\ i \\ \hline \end{array}$

5. $\begin{array}{r} 14 \\ \times\ L \\ \hline \end{array}$

6. $\begin{array}{r} 13 \\ \times\ Q \\ \hline \end{array}$

7. $\begin{array}{r} 18 \\ \times\ V \\ \hline \end{array}$

8. $\begin{array}{r} 12 \\ \times\ L \\ \hline \end{array}$

9. $\begin{array}{r} 11 \\ \times\ Y \\ \hline \end{array}$

10. $\begin{array}{r} 10 \\ \times\ O \\ \hline \end{array}$

# MULTIPLICATION - 2

MULTIPLYING SINGLE DIGIT NUMBERS
THEN SUBTRACTING A NUMBER TO EQUAL A LETTER

1. $8 \times 3 - 4 =$
2. $6 \times 4 - 6 =$
3. $7 \times 4 - 9 =$
4. $5 \times 4 - 7 =$
5. $9 \times 2 - 3 =$
6. $4 \times 3 - 5 =$
7. $3 \times 3 - 2 =$
8. $2 \times 2 - 1 =$

# MULTIPLICATION - 3

MULTIPLYING SINGLE DIGIT NUMBERS
THEN SUBTRACTING A NUMBER TO EQUAL A LETTER

1. $8 \times 6 - 30 =$
2. $7 \times 5 - 27 =$
3. $6 \times 4 - 21 =$
4. $7 \times 6 - 37 =$
5. $6 \times 6 - 24 =$
6. $9 \times 4 - 23 =$
7. $7 \times 8 - 48 =$
8. $8 \times 2 - 4 =$

# MULTIPLICATION - 4

MULTIPLYING LETTERS AND NUMBERS
THEN ADDING A LETTER TO EQUAL A NUMBER

1.

```
   P
 x 6
-----
=
 + G
-----
```

2.

```
   S
 x 5
-----
=
 + K
-----
```

3.

```
   Z
 x 7
-----
=
 + O
-----
```

4.

```
   V
 x 9
-----
=
 + T
-----
```

5.

```
   M
 x 2
-----
=
 + F
-----
```

6.

```
   Y
 x 8
-----
=
 + N
-----
```

7.

```
   X
 x 3
-----
=
 + E
-----
```

8.

```
   L
 x 4
-----
=
 + i
-----
```

# MULTIPLICATION - 4

## MULTIPLYING LETTERS AND NUMBERS
## THEN ADDING A LETTER TO EQUAL A NUMBER
## ANSWER KEY

1. P
x 6
96
+ G
103

2. S
x 5
95
+ K
106

3. Z
x 7
182
+ O
197

4. V
x 9
198
+ T
218

5. M
x 2
26
+ F
32

6. Y
x 8
200
+ N
214

7. X
x 3
72
+ E
77

8. L
x 4
48
+ i
57

# MULTIPLICATION - 4

MULTIPLYING NUMBERS
THEN SUBTRACTING NUMBERS
TO EQUAL A LETTER

1. $12 \times 10 = \_\_\_$, $- 99 = \_\_\_$

2. $75 \times 2 = \_\_\_$, $- 139 = \_\_\_$

3. $66 \times 10 = \_\_\_$, $- 638 = \_\_\_$

4. $11 \times 13 = \_\_\_$, $- 133 = \_\_\_$

5. $80 \times 10 = \_\_\_$, $- 790 = \_\_\_$

6. $50 \times 3 = \_\_\_$, $- 141 = \_\_\_$

7. $64 \times 2 = \_\_\_$, $- 117 = \_\_\_$

8. $53 \times 4 = \_\_\_$, $- 201 = \_\_\_$

# MULTIPLICATION - 5

## MULTIPLYING NUMBERS THEN SUBTRACTING LETTERS

1.
$$\begin{array}{r} 20 \\ \times\ 12 \\ \hline = \\ -\ D \\ \hline \end{array}$$

2.
$$\begin{array}{r} 18 \\ \times\ 16 \\ \hline = \\ -\ X \\ \hline \end{array}$$

3.
$$\begin{array}{r} 17 \\ \times\ 14 \\ \hline = \\ -\ G \\ \hline \end{array}$$

4.
$$\begin{array}{r} 19 \\ \times\ 15 \\ \hline = \\ -\ M \\ \hline \end{array}$$

5.
$$\begin{array}{r} 11 \\ \times\ 10 \\ \hline = \\ -\ O \\ \hline \end{array}$$

6.
$$\begin{array}{r} 13 \\ \times\ 12 \\ \hline = \\ -\ L \\ \hline \end{array}$$

7.
$$\begin{array}{r} 12 \\ \times\ 11 \\ \hline = \\ -\ O \\ \hline \end{array}$$

8.
$$\begin{array}{r} 16 \\ \times\ 13 \\ \hline = \\ -\ T \\ \hline \end{array}$$

# MULTIPLICATION - 6

## MULTIPLYING LETTERS THEN SUBTRACTING NUMBERS < 26

1.
J
x K
=
- 24

2.
Z
x F
=
- 19

3.
N
x M
=
- 17

4.
S
x R
=
- 18

5.
Q
x i
=
= 25

6.
W
x V
=
- 22

7.
U
x P
=
- 16

8.
X
x Y
=
- 20

# MULTIPLICATION - 7

## MULTIPLYING LETTERS THEN SUBTRACTING LETTERS TO EQUAL A NUMBER

1.
```
     M
   x N
______
=
   - P
______
```

2.
```
     J
   x K
______
=
   - F
______
```

3.
```
     L
   x P
______
=
   - E
______
```

4.
```
     P
   x G
______
=
   - H
______
```

5.
```
     T
   x V
______
=
   = O
______
```

6.
```
     I
   x H
______
=
   - G
______
```

7.
```
     R
   x S
______
=
   - B
______
```

8.
```
     U
   x M
______
=
   - I
______
```

# MULTIPLICATION - 8

## MULTIPLYING NUMBERS THEN FIND THE NUMBER TO SUBTRACT TO EQUAL THE LETTER

1. 12 x 10 = ___ - ? = M

2. 11 x 10 = ___ - ? = N

3. 13 x 10 = ___ - ? = E

4. 14 x 8 = ___ - ? = L

5. 15 x 9 = ___ - ? = O

6. 16 x 10 = ___ - ? = Q

7. 16 x 11 = ___ - ? = S

8. 17 x 12 = ___ - ? = Y

# MULTIPLICATION - 9

## MULTIPLYING LETTERS THEN MULTIPLYING NUMBERS

1.
```
   M
 x L
----
=
 x 9
----
```

2.
```
   O
 x i
----
=
 x 7
----
```

3.
```
   Q
 x J
----
=
 x 6
----
```

4.
```
   Y
 x X
----
=
 x 8
----
```

5.
```
   Z
 x K
----
=
 x 3
----
```

6.
```
   T
 x G
----
=
 x 5
----
```

7.
```
   V
 x R
----
=
 x 6
----
```

8.
```
   U
 x P
----
=
 x 2
----
```

# MULTIPLICATION - 10

## MULTIPLYING LETTERS THEN ADDING NUMBERS THEN SUBTRACTING NUMBERS TO EQUAL A LETTER

1. 8 x 5 + 9 – 23 =
2. 5 x 5 + 1 – 15 =
3. 6 x 6 + 4 – 19 =
4. 7 x 3 + 8 – 7 =
5. 6 x 5 + 6 – 14 =
6. 3 x 3 + 9 – 8 =
7. 5 x 3 + 11 – 12 =
8. 7 x 7 + 8 – 34 =

UNIT 6

# DIVISION

# DIVISION – 1

## MULTIPLYING LETTERS WITH LETTERS THEN DIVIDING THE NUMBER BY 4 WITH ANY REMAINDER A LETTER

1. K
   x J
   =
   ÷ 4

2. S
   x Q
   =
   ÷ 4

3. E
   x D
   =
   ÷ 4

4. U
   x B
   =
   ÷ 4

5. X
   x G
   =
   ÷ 4

6. W
   x A
   =
   ÷ 4

7. O
   x i
   =
   ÷ 4

8. M
   x J
   =
   ÷ 4

# DIVISION – 2

## MULTIPLYING NUMBERS < 51
## THEN DIVIDING BY LETTERS TO EQUAL A NUMBER

1. 50
x 33
=
÷ J

2. 47
x 26
=
÷ Z

3. 49
x 41
=
÷ Q

4. 48
x 43
=
÷ K

5. 50
x 42
=
÷ S

6. 45
x 42
=
÷ T

7. 46
x 38
=
÷ V

8. 44
x 44
=
÷ H

# DIVISION - 3

DIVIDING NUMBERS TO EQUAL A LETTER
THEN COMBINE THE LETTERS TO MAKE A WORD

1. 20÷10= 10÷10= 200÷10= WORD =
2. 50÷10= 50÷50= 180÷10= WORD =
3. 60÷10= 60÷60= 200÷10= WORD =
4. 110÷10= 90÷10 140÷10= 70÷10= WORD =
5. 150÷10= 160÷10= 50÷10= 140÷10= WORD =
6. 9÷3= 9÷9= 20÷10= WORD =
7. 100÷10= 100÷100= 900÷100= 120÷10= WORD =
8. 160÷10= 500÷100= 140÷10= WORD =

# UNIT 7

# EXTRA CHALLENGE

# EXTRA CHALLENGE – 1

WHAT DO THESE WORDS HAVE IN COMMON?
(WHAT IS THEIR NAME OR NAMES)

1. CAT, DOG, BIRD
2. HOUSE, MOBILE HOME, APARTMENT
3. FIRE DEPARMENT, POLICE DEPARTMENT, AMBULANCE
4. OCEAN, POND, BEACH, RIVER
5. SEAGULL, HAWK, DUCK
6. RED, BLUE, ORANGE, GRAY, BLACK

UNIT 8

# ANSWER KEY

# NUMBER TO WORDS - 1

## ANSWER KEY

1. 7 9 18 12 19 = GIRLS
2. 11 9 14 4 = KIND
3. 2 15 25 19 = BOYS
4. 17 21 5 5 14 19 = QUEENS
5. 13 1 14 1 7 5 = MANAGE
6. 16 1 18 5 14 20 19 = PARENTS
7. 6 18 9 4 7 5 = FRIDGE
8. 22 5 8 9 3 12 5 = VEHICLE

# NUMBERS TO LETTERS - 2

ANSWER KEY

1. 6 9 7 8 20 5 18 = FIGHTER
2. 3 1 18 19 = CARS
3. 16 12 1 14 5 19 = PLANES
4. 1 4 15 12 5 19 3 5 14 20 = ADOLESCENT
5. 19 20 21 4 5 14 20 = STUDENT
6. 7 18 1 14 4 11 9 4 = GRANDKID
7. 11 9 14 7 = KING
8. 17 21 5 5 14 = QUEEN

# NUMBERS TO WORDS - 3

## ANSWER KEY

1. 4 5 1 20 8 = DEATH
2. 8 5 1 20 = HEAT
3. 12 5 6 20 = LEFT
4. 13 15 14 5 25 = MONEY
5. 16 15 15 18 = POOR
6. 12 15 1 14 = LOAN
7. 6 1 13 9 12 25 = FAMILY

# NUMBERS TO WORDS - 4

## ALL END WITH AN "S"
## ANSWER KEY

1. 2 1 20 19 = BATS
2. 11 9 4 19 = KIDS
3. 12 5 7 19 = LEGS
4. 16 1 18 5 14 20 19 = PARENTS
5. 19 20 9 3 11 19 = STICKS
6. 18 5 16 20 9 12 5 19 = REPTILES
7. 19 9 2 12 9 14 7 19 = SIBLINGS

## NUMBERS TO WORDS - 5

ALL END WITH "ED"
ANSWER KEY

1. 13 1 18 18 9 5 4 = MARRIED
2. 16 12 1 14 20 5 4 = PLANTED
3. 7 18 1 14 20 5 4 = GRANTED
4. 3 15 22 5 18 5 4 = COVERED
5. 23 15 18 18 9 5 4 = WORRIED
6. 20 18 9 5 4 = TRIED
7. 15 16 5 14 5 4 = OPENED

# NUMBERS TO WORDS - 6

STATES
ANSWER KEY

1. 3 15 12 15 18 1 4 15 = COLORADO
2. 3 15 14 14 5 3 20 9 3 21 20 = CONNECTICUT
3. 23 25 15 13 9 14 7 = WYOMING
4. 13 9 3 8 9 7 1 14 = MICHIGAN
5. 7 5 15 18 7 9 1 = GEORGIA
6. 1 12 1 2 1 13 1 = ALABAMA
7. 14 5 2 18 1 19 11 1 = NEBRASKA

# NUMBERS TO WORDS - 7

BOYS NAMES
ANSWER KEY

1. 2 18 9 1 14 = BRIAN
2. 2 18 1 4 = BRAD
3. 6 18 5 4 = FRED
4. 10 15 8 14 = JOHN
5. 19 20 5 22 5 = STEVE
6. 4 1 22 9 4 = DAVID
7. 13 9 11 5 = MIKE
8. 3 8 18 9 19 = CHRIS

# NUMBERS TO WORDS - 8

## GIRLS NAMES
## ANSWER KEY

1. 12 9 19 1 = LISA
2. 10 5 14 = JEN
3. 16 1 13 = PAM
4. 4 5 2 2 9 5 = DEBBIE
5. 2 18 9 4 7 5 20 = BRIDGET
6. 19 20 5 16 8 1 14 9 5 = STEPHANIE
7. 10 5 19 19 9 3 1 = JESSICA

# ADDITION – 1

ADDING NUMBERS TO EQUAL A LETTER
NO LETTER WILL BE THE SAME
ANSWER KEY

1. 9+10 = 19 = S
2. 6+15 = 21 = U
3. 1+13 = 14 = N
4. 22+2 = 24 = X
5. 4+8 = 12 = L
6. 11+5 = 16 = P
7. 3+14 = 17 = Q

# ADDITION - 2

ADDING NUMBERS TO EQUAL A LETTER
THEN COMBINE THE LETTERS
TO MAKE A 7 LETTER WORD
ANSWER KEY

1. 15+4 = 19 = S
2. 2+1 = 3 = C
3. 4+4 = 8 = H
4. 11+4 = 15 = O
5. 9+6 = 15 = O
6. 6+6 = 12 = L
7. 11+8 = 19 = S

ANSWER: SCHOOLS

# ADDITION - 3

ADDING NUMBERS TO EQUAL A LETTER
THEN COMBINE THE LETTERS
TO MAKE AN 8 LETTER WORD
ANSWER KEY

1. 10+10=20=T
2. 3+2=5=E
3. 0+1=1=A
4. 2+1=3=C
5. 4+4=8=H
6. 4+1=5=E
7. 10+8=18=R
8. 10+9=19=S

ANSWER: TEACHERS

# ADDITION - 4

## ADDING NUMBERS TO MAKE A WORD

1. 3+10 2+13 0+1 12+2 6+3 13+1 2+5 =
   13 15 1 14 9 14 7 = MOANING

2. 4+2 1+4 11+2 0+1 10+2 1+4 16+3 =
   6 5 13 1 12 5 19 = FEMALES

3. 16+2 3+2 17+2 18+2 3+2 2+2 =
   18 5 19 20 5 4 = RESTED

4. 16+3 15+5 17+4 0+4 1+4 12+2 18+2 =
   19 20 21 4 5 14 20 = STUDENT

5. 9+2 8+1 17+3 10+10 3+2 11+3 =
   11 9 20 20 5 14 = KITTEN

6. 11+9 6+3 4+3 4+1 16+2 =
   20 9 7 5 18 = TIGER

7. 14+3 18+3 0+1 10+2 6+3 18+2 12+13 =
   17 21 1 12 9 20 25 = QUALITY

# ADDITION - 5

## ADDING LETTERS TO EQUAL A NUMBER
## ANSWER KEY

1. B + M + H + I = 2 + 13 + 8 + 9 = 32
2. Q + T + S + K = 17 + 20 + 19 + 11 = 67
3. A + M + D + G = 1 + 13 + 4 + 7 = 25
4. S + U + M + O = 19 + 21 + 13 + 15 = 68
5. X + W + Y + Z = 24 + 23 + 25 + 26 = 98
6. Y + R + B + G = 25 + 18 + 2 + 7 = 52
7. V + Q + i + W = 22 + 17 + 9 + 23 = 71
8. Z + U + T + P = 26 + 21 + 20 + 16 = 83
9. U + H + M + O = 21 + 8 + 13 + 15 = 57
10. V + i + J + M = 22 + 9 + 10 + 13 = 54

# ADDITION - 6

## ADDING LETTERS IN A WORD TO EQUAL A NUMBER ANSWER KEY

1. KITTEN = 11+9+20+20+5+14 = 79
2. TEACHER = 20+5+1+3+8+5+18 = 60
3. PARENT = 16+1+18+5+14+20 = 74
4. HOUSE = 8+15+21+19+5 = 68
5. COUNTRY = 3+15+21+14+20+18+25 = 116
6. STATE = 19+20+1+20+5 = 65
7. COUNTY = 3+15+21+14+20+25 = 98
8. ANIMAL = 1+14+9+13+1+12 = 50

# SUBTRACTION - 1

SUBTRACTING NUMBERS TO GET A LETTER
NO LETTER WILL BE THE SAME
ANSWER KEY

1. 36-28 = 8 = H
2. 44-20 = 24 = X
3. 54-34 = 20 =T
4. 55-43 = 12 = L
5. 60-58 = 2 = B
6. 53-28 = 25 = Y
7. 61-60 = 1 = A

# SUBTRACTION - 2

## SUBTRACTING NUMBERS TO MAKE A WORD
## ANSWER KEY

1. 5-2 4-3 21-7 6-2 24-12 15-10 =
   3 1 14 4 12 5 = CANDLE

2. 12-8 18-3 20-8 24-12 25-6 =
   4 25 23 23 29 = DOLLS

3. 17-5 11-2 23-4 22-2 =
   12 9 19 20 = LIST

4. 26-7 21-1 12-7 21-5 22-3 =
   19 20 5 16 19 = STEPS

5. 12-8 11-6 6-5 16-10 =
   4 5 1 6 = DEAF

6. 15-3 15-6 17-10 10-2 22-2 22-3 =
   12 9 7 8 20 = LIGHTS

7. 15-4 14-5 7-3 24-5 =
   11 9 4 19 = KIDS

# SUBTRACTION - 3

SUBRACTING LETTERS FROM NUMBERS
THE ANSWER IS A LETTER
ANSWER KEY

1. 32 - J = 22 = V

2. 29 - Y = 4 = D

3. 27 - F = 21 = U

4. 33 - V = 11 = K

5. 30 - T = 10 = J

6. 37 - S = 18 = R

7. 31 - X = 7 = G

8. 34 - S = 15 = O

9. 35 - Q = 18 = R

10. 40 - U = 19 = S

# ADDITION WITH SUBTRACTION - 1

## ADDING SINGLE DIGIT NUMBERS THEN SUBTRACTING SINGLE DIGIT NUMBERS TO EQUAL A LETTER ANSWER KEY

1.
$$\begin{array}{r} 3 \\ +\ 5 \\ \hline 8 \\ -\ 4 \\ \hline 4 = D \end{array}$$

2.
$$\begin{array}{r} 5 \\ +\ 8 \\ \hline 13 \\ -\ 3 \\ \hline 10 = J \end{array}$$

3.
$$\begin{array}{r} 7 \\ +\ 9 \\ \hline 16 \\ -\ 6 \\ \hline 10 = J \end{array}$$

4.
$$\begin{array}{r} 9 \\ +\ 9 \\ \hline 18 \\ -\ 7 \\ \hline 11 = K \end{array}$$

5.
$$\begin{array}{r} 8 \\ +\ 6 \\ \hline 14 \\ -\ 2 \\ \hline 12 = L \end{array}$$

6.
$$\begin{array}{r} 7 \\ +\ 7 \\ \hline 14 \\ -\ 5 \\ \hline 9 = I \end{array}$$

7.
$$\begin{array}{r} 4 \\ +\ 3 \\ \hline 7 \\ -\ 1 \\ \hline 6 = F \end{array}$$

8.
$$\begin{array}{r} 6 \\ +\ 6 \\ \hline 12 \\ -\ 8 \\ \hline 4 = D \end{array}$$

# ADDITION WITH SUBTRACTION - 2

ADDING NUMBERS THEN
SUBTRACTING NUMBERS TO EQUAL A LETTER
ANSWER KEY

1.

$$\begin{array}{r} 10 \\ +\ 4 \\ \hline 14 \\ -\ 3 \\ \hline 11=K \end{array}$$

2.

$$\begin{array}{r} 13 \\ +\ 9 \\ \hline 22 \\ -\ 5 \\ \hline 17=Q \end{array}$$

3.

$$\begin{array}{r} 18 \\ +\ 7 \\ \hline 25 \\ -\ 5 \\ \hline 20=T \end{array}$$

4.

$$\begin{array}{r} 24 \\ +\ 9 \\ \hline 33 \\ -\ 7 \\ \hline 26=Z \end{array}$$

5.

$$\begin{array}{r} 21 \\ +\ 7 \\ \hline 28 \\ -\ 8 \\ \hline 20=T \end{array}$$

6.

$$\begin{array}{r} 16 \\ +\ 11 \\ \hline 27 \\ -\ 4 \\ \hline 23=W \end{array}$$

7.

$$\begin{array}{r} 15 \\ +\ 14 \\ \hline 29 \\ -\ 13 \\ \hline 16=P \end{array}$$

# ADDITION WITH SUBTRACTION - 3

ADDING NUMBERS THEN SUBTRACTING NUMBERS
THE ANSWER IS A LETTER
ANSWER KEY

1.
  31
+ 29
  60
- 37
23 = W

2.
  34
+ 26
  60
- 38
22 = V

3.
  38
+ 29
  67
- 47
20 = T

4.
  23
+21
  44
- 35
9 = i

5.
  33
+ 31
  64
- 51
13 = M

6.
  14
+ 12
  26
- 18
8 = H

7.
  47
+ 42
  89
- 82
7 = G

8.
  49
+ 32
  81
- 77
4 = D

# ADDITION WITH SUBTRACTION - 4

ADDING NUMBERS THEN SUBTRACTING NUMBERS
THE ANSWER IS A LETTER
ANSWER KEY

1.
36
+ 33
= 69
- 56
13=M

2.
47
+ 23
= 70
- 49
21=U

3.
39
+ 17
= 56
- 44
12=L

4.
58
+ 28
= 86
- 67
19=S

5.
64
+ 43
= 107
- 97
10=J

6.
59
+ 31
= 90
- 79
11=K

7.
64
+ 22
= 86
- 84
2=B

8.
55
+ 21
= 76
- 67
9=i

# ADDITION WITH SUBTRACTION - 5

ADDING NUMBERS THEN SUBRACTING LETTERS
ANSWER KEY

1.
25
+ 22
47
- P
31

2.
28
+ 24
52
- H
44

3.
29
+ 21
50
- K
39

4.
32
+ 30
612
- R
44

5.
37
+ 26
63
- V
41

6.
39
+ 35
74
- Y
49

7.
33
+ 28
61
- B
59

8.
40
+ 34
74
- G
67

# ADDITION WITH SUBTRACTION - 6

## ADDING LETTERS THEN SUBTRACTING NUMBERS
## ANSWER KEY

1.
W
+ G
30
- 7
23

2.
N
+ H
22
- 10
12

3.
P
+ O
31
- 8
23

4.
S
+ Q
36
- 5
31

5.
X
+ K
35
- 12
23

6.
Z
+ i
35
- 3
32

7.
S
+ F
25
- 4
21

8.
V
+ M
35
- 9
26

# ADDITION WITH SUBTRACTION – 7

ADDING NUMBERS AND LETTERS
THEN SUBTRACTING WITH LETTERS
ANSWER KEY

1.

28
+ Z
54
- E
49

2.

31
+ R
49
- G
42

3.

33
+ Y
58
- D
54

4.

37
+ S
56
- K
45

5.

39
+ M
52
- V
30

6.

41
+ i
50
- Y
25

7.

44
+ Q
61
- T
41

8.

42
+ N
56
- W
33

# ADDITION WITH SUBTRACTION - 8

## ADDING LETTERS
## THEN SUBTRACTING LETTERS
## TO EQUAL A NUMBER
## ANSWER KEY

1. T + M - K = 20+13=33-11=22
2. Q + L - N = 17+12=29-14=15
3. Z + X - U = 26+24=50-21=29
4. J + I - D = 10+9=19-4=15
5. V + W - H = 22+23=45-8=37
6. P + O - B = 16+15=31-2=29
7. R + G - C = 18+7=25-3=22
8. U + J – E = 21+10=31-5=26

# ADDITION WITH SUBTRACTION - 9

## ADDING THE LETTERS IN A WORD THEN SUBTRACTING A NUMBER TO EQUAL A LETTER ANSWER KEY

1. HIGH – 12 = 8+9+7+8=32-12=20=T
2. CALM – 11 = 3+1+12+13=29-11=18=R
3. KIND – 19 = 11+9+14+4=38-19=19=S
4. LOVE – 32 = 12+15+22+5=54-32=22=V
5. JOBS – 39 = 10+15+2+19=46-39=7=G
6. BRAVE – 44 = 2+18+1+22+5=48-44=4=D
7. SWEET – 63 = 19+23+5+5+20=72-63=9=i
8. SCHOOL - 61 = 19+3+8+15+15+12=72-61=11-K

# MULTIPLYING – 1

## NUMBERS MULTIPLYING LETTERS
## NUMBERS 20 AND UNDER
## ANSWER KEY

1. 19 x Y = 475

2. 17 x F = 102

3. 16 x Z = 416

4. 20 x i = 180

5. 14 x L = 168

6. 13 x Q = 221

7. 18 x V = 396

8. 12 x L = 144

9. 11 x Y = 275

10. 10 x O = 150

# MULTIPLICATION - 2

MULTIPLYING SINGLE DIGIT NUMBERS
THEN SUBTRACTING A NUMBER TO EQUAL A LETTER
ANSWER KEY

1. 8 x 3 = 24 – 4 = 20 = T
2. 6 x 4 = 24 – 6 = 18 = R
3. 7 x 4 = 28 – 9 = 19 = S
4. 5 x 4 = 20 – 7 = 13 = M
5. 9 x 2 = 18 – 3 = 15 = O
6. 4 x 3 = 12 – 5 = 7 = G
7. 3 x 3 = 9 – 2 = 7 = G
8. 2 x 2 = 4 – 1 = 3 = C

# MULTIPLICATION - 3

MULTIPLYING SINGLE DIGIT NUMBERS
THEN SUBTRACTING A NUMBER TO EQUAL A LETTER
ANSWER KEY

1. $8 \times 6 = 48 - 30 = 18$ = R
2. $7 \times 5 = 35 - 27 = 8$ = H
3. $6 \times 4 = 24 - 21 = 3$ = C
4. $7 \times 6 - 37 = 42 - 37 = 5$ = E
5. $6 \times 6 - 24 = 36 - 24 = 12$ = L
6. $9 \times 4 = 36 - 23 = 13$ = M
7. $7 \times 8 = 56 - 48 = 8$ = H
8. $8 \times 2 = 16 - 4 = 12$ = L

# MULTIPLICATION - 4

## MULTIPLYING NUMBERS THEN SUBTRACTING NUMBERS TO EQUAL A LETTER ANSWER KEY

1. 12
x 10
= 120
- 99
21=U

2. 75
x 2
= 150
- 139
11=K

3. 66
x 10
= 660
- 638
22=V

4. 11
x 13
= 143
- 133
10=J

5. 80
x 10
= 800
- 790
10=J

6. 50
x 3
= 150
- 141
9=i

7. 64
x 2
= 128
- 117
11=K

8. 53
x 4
= 212
- 201
11=K

# MULTIPLICATION - 5

## MULTIPLYING NUMBERS THEN SUBTRACTING LETTERS ANSWER KEY

1.
```
  20
x 12
----
 240
 - D
----
 236
```

2.
```
  18
x 16
----
 288
 - X
----
 264
```

3.
```
  17
x 14
----
 238
 - G
----
 231
```

4.
```
  19
x 15
----
 285
 - M
----
 272
```

5.
```
  11
x 10
----
 110
 - O
----
  95
```

6.
```
  13
x 12
----
 156
 - L
----
 144
```

7.
```
  12
x 11
----
 132
 - O
----
 117
```

8.
```
  16
x 13
----
 208
 - T
----
 188
```

# MULTIPLICATION - 6

## MULTIPLYING LETTERS THEN SUBTRACTING NUMBERS < 26 ANSWER KEY

1\.

| |
|---:|
| J |
| x K |
| 110 |
| - 24 |
| 86 |

2\.

| |
|---:|
| Z |
| x F |
| 156 |
| - 19 |
| 137 |

3\.

| |
|---:|
| N |
| x M |
| 182 |
| - 17 |
| 165 |

4\.

| |
|---:|
| S |
| x R |
| 342 |
| - 18 |
| 324 |

5\.

| |
|---:|
| Q |
| x i |
| 153 |
| = 25 |
| 128 |

6\.

| |
|---:|
| W |
| x V |
| 506 |
| - 22 |
| 484 |

7\.

| |
|---:|
| U |
| x P |
| 336 |
| - 16 |
| 320 |

8\.

| |
|---:|
| X |
| x Y |
| 600 |
| - 20 |
| 580 |

# MULTIPLICATION - 7

## MULTIPLYING LETTERS THEN SUBTRACTING LETTERS TO EQUAL A NUMBER ANSWER KEY

1.
M=13
x N=14
182
- P=16
166

2.
J=10
x K=11
110
- F=6
104

3.
L=12
x P=16
192
- E=5
187

4.
P=16
x G=7
112
- H=8
104

5.
T=20
x V=22
440
- O=15
425

6.
I=9
x H=8
72
- G=7
65

7.
R=18
x S=19
342
- B=2
340

8.
U=21
x M=13
273
- I=9
264

# MULTIPLICATION - 8

MULTIPLYING NUMBERS THEN
FIND THE NUMBER TO SUBTRACT
TO EQUAL THE LETTER
ANSWER KEY

1.
$$\begin{array}{r} 12 \\ \times\ 10 \\ \hline 120 \\ -\ 107 \\ \hline M=13 \end{array}$$

2.
$$\begin{array}{r} 11 \\ \times\ 10 \\ \hline 110 \\ -\ 96 \\ \hline N=14 \end{array}$$

3.
$$\begin{array}{r} 13 \\ \times\ 10 \\ \hline 130 \\ -\ 125 \\ \hline E=5 \end{array}$$

4.
$$\begin{array}{r} 14 \\ \times\ 8 \\ \hline 112 \\ -\ 100 \\ \hline L=12 \end{array}$$

5.
$$\begin{array}{r} 15 \\ \times\ 9 \\ \hline 135 \\ -\ 120 \\ \hline O=15 \end{array}$$

6.
$$\begin{array}{r} 16 \\ \times\ 10 \\ \hline 160 \\ -\ 143 \\ \hline Q=17 \end{array}$$

7.
$$\begin{array}{r} 16 \\ \times\ 11 \\ \hline 176 \\ -\ 157 \\ \hline S=19 \end{array}$$

8.
$$\begin{array}{r} 17 \\ \times\ 12 \\ \hline 204 \\ -\ 179 \\ \hline Y=25 \end{array}$$

# MULTIPLICATION - 9

## MULTIPLYING LETTERS THEN MULTIPLYING NUMBERS ANSWER KEY

1.
M
x L
156
x 9
1404

2.
O
x i
135
x 7
945

3.
Q
x J
170
x 6
54
1020

4.
Y
x X
600
x 8
4800

5.
Z
x K
286
x 3
858

6.
T
x G
140
x 5
25
700

7.
V
x R
396
x 6
2376

8.
U
x P
336
x 2
672

# MULTIPLICATION - 10

## MULTIPLYING LETTERS THEN ADDING NUMBERS THEN SUBTRACTING NUMBERS TO EQUAL A LETTER ANSWER KEY

1. 8 x 5 + 9 – 23 = 8 x 5 = 40 + 9 = 49 – 23 = 26 = Z
2. 5 x 5 + 1 – 15 = 5 x 5 = 25 + 1 = 26 – 15 = 11 = K
3. 6 x 6 + 4 – 19 = 6 x 6 = 36 + 4 = 40 – 19 = 21 = U
4. 7 x 3 + 8 – 7 = 7 x 3 = 21 + 8 = 29 – 7 = 22 = V
5. 6 x 5 + 6 – 14 = 6 x 5 = 30 + 6 = 36 – 14 = 22 = V
6. 3 x 3 + 9 – 8 = 3 x 3 = 9 + 9 = 18 – 8 = 10 = J
7. 5 x 3 + 11 – 12 = 5 x 3 = 15 + 11 = 26 – 12 = 14 = N
8. 7 x 7 + 8 – 34 = 7 x 7 = 49 + 8 = 57 – 34 = 23 = W

# DIVISION - 1

## MULTIPLYING LETTERS WITH LETTERS
## THEN DIVIDING THE NUMBER BY 4
## WITH THE REMAINDER A LETTER
## ANSWER KEY

1. K
   x J
   110
   ÷ 4
   27
   r 2 = B

2. S
   x Q
   323
   ÷ 4
   80
   r 3 = C

3. E
   x D
   20
   ÷ 4
   5
   r = 0

4. U
   x B
   42
   ÷ 4
   10
   r 2 = B

5. X
   x G
   168
   ÷ 4
   42
   r = 0

6. W
   x A
   23
   ÷ 4
   5
   r 3 = C

7. O
   x i
   135
   ÷ 4
   33
   r 3 = C

8. M
   x J
   130
   ÷ 4
   32
   r 2 = B

# DIVISION - 2

## MULTIPLYING NUMBERS < 51
## THEN DIVIDING BY A LETTER TO EQUAL A NUMBER
## ANSWER KEY

1.
50
x 33
1650
÷ J
165

2.
47
x 26
1222
÷ Z
47

3.
49
x 41
2009
÷ Q
118
r 3

4.
48
x 43
2064
÷ K
187
r 8

5.
50
x 42
2100
÷ S
110
r 10

6.
45
x 42
1890
÷ T
94
r 10

7.
46
x 38
1748
÷ V
79
r 10

8.
44
x 44
1936
÷ H
242

# DIVISION - 3

## DIVIDING NUMBERS TO EQUAL A LETTER THEN COMBINE THE LETTERS TO MAKE A WORD ANSWER KEY

1. 20÷10=2=B 10÷10=1=A 200÷10=20=7 WORD = BAT
2. 50÷10=5=E 50÷50=1=A 180÷10=18=R WORD = EAR
3. 60÷10=6=F 60÷60=1=A 200÷10=20=T WORD = FAT
4. 110÷10=11=K 90÷10=9=I 140÷10=14=N 70÷10=7=G

   WORD = KING
5. 150÷10=15=O 160÷10=16=P 50÷10=5=E 140÷10=14=N

   WORD = OPEN
6. 9÷3=3=C 9÷9=1=A 20÷10=2=B WORD = CAB
7. 100÷10=10=J 100÷100=1=A 900÷100=9=I

   120÷10=12=L WORD = JAIL
8. 160÷10=16=P 500÷100=5=E 140÷10=14=N

   WORD = PEN

# EXTRA CHALLENGE – 1

WHAT DO THESE WORDS HAVE IN COMMON?
(WHAT IS THEIR NAME OR NAMES)
ANSWER KEY

1. CAT, DOG, BIRD = PETS OR ANIMALS
2. HOUSE, MOBILE HOME, APARTMENT = PLACES YOU LIVE
3. FIRE DEPARMENT, POLICE DEPARTMENT, AMBULANCE = ALL RELATED TO 911 EMERGENCY SERVICES
4. OCEAN, POND, BEACH, RIVER = WATER
5. SEAGULL, HAWK, DUCK = ANIMALS OR BIRDS
6. RED, BLUE, ORANGE, GRAY, BLACK = COLORS

Printed in the United States
by Baker & Taylor Publisher Services